YOUR KNOWLEDGE HAS VALUE

- We will publish your bachelor's and master's thesis, essays and papers

- Your own eBook and book - sold worldwide in all relevant shops

- Earn money with each sale

Upload your text at www.GRIN.com and publish for free

Bibliographic information published by the German National Library:

The German National Library lists this publication in the National Bibliography; detailed bibliographic data are available on the Internet at http://dnb.dnb.de .

Imprint:

Copyright © 2018 GRIN Verlag
Print and binding: Books on Demand GmbH, Norderstedt Germany
ISBN: 9783668940123

This book at GRIN:

https://www.grin.com/document/455533

Shedrach Ugwu

Soil Quality Improvement of Acidic Soil and Its Effects on The Growth and Yield of Maize

Using Biochar Derived From Rice Waste

GRIN Verlag

Table of Contents

1. INTRODUCTION

It has been reported that poor plant growth yields are strongly associated with low pH because of the deficiencies and toxicities related to a number of elements (Blamey and Chapman, 1982). Also, acidic soil with (pH <0.5) is a wide spread problem in Agbani Area South East Nigeria which has had adverse effect including decreased soil productivity and sub optimal plant growth. These acidic soils originates from acidic paren materials, mainly granite and granite gnesis that are naturally more acidic compared to the soil generated from calcerous shale or lime stone.

Moreover, atmospheric pollutant of industrial origin can increase the acidic content of soils. A gradual increase in acid deposition has occurred as a result of heavy industrialization since the 1970s. Maize belongs to the grains under the family of graminae and class of cereals. It thrives under a wide range of environmental conditions for optimum production they require a lot of sunshine, warmth and an average temperature below 14°C and a well distributed amount of rainfall. It prefers a pH of 6.5-8.0 (IITA, 1982) while a strongly acid soil is unsuitable for good yield. The soil must be deep, well drained, structured with adequate organic matter and inorganic matter like weathered minerals within the rooting depth. Biochar is a charcoal used as a soil ameliorant. It is recognized as a multifunctional material that can be used in carbon sequestration, metal immobilization and soil fertility increase. Biochar is a stable solid, rich in carbon and can endure in the soil for thousands of years. Like most charcoal, biochar is made from renewable materials via pysolysis. The alkaline nature of biochar along with the presence of a variety of other elements depending on its origin makes it a good material as a soil conditioner.

Soil quality is the ability of a soil to perform functions that are essential to plant growth and the environment. It is a reflection of how well a soil performs the functions of maintaining biodiversity and productivity. Soil quality improvement using biochar not only increases soil pH level but also affect the availability of other minerals and nutrients found in the soil (Mclean 1982). To improve the quality of an acidic soil with respect to pH, the application of biochar as an ameliorant to acidic soils is seldom. It is affordable and cost effective making it an alternative.

The main objective of this study is to use biochar derived from rice mill waste to ameliorate acidic soil and its effect on the growth and yield of maize.

2. LITERATURE REVIEW

2.1. Maize: General Description

Maize is a grass and it belongs to the large family called Gramineae, subfamily Andropogodeae and class of cereals. The origin of maize is still a matter of speculation because no wild forms of maize have been found. A current research showed that maize had its origin from a primitive type of corn, which have been found in caves or dwelling of early man about 7000 years old. However, it was introduced into West Africa by the Portuguese.

It is gotten throughout the tropics and in the temperate regions of the world. Maize has a fibrous root system which grows mainly from the lower nodes of the stem below ground level. There are also seminal and prop root, the prop or brace roots are adventitious roots that rise from nodes that are above the soil surface. In Nigeria, it is cultivated mostly in the southern part of the country.

Maize plant is an annual plant, the stem, culms or stalks are solid unbranched and herbaceous. Occasionally, ear bearing branches are formed. Maize stems range in height from 0.6 to 6 m and are grooved on alternating sides of each internodes. The internodes are shorter towards the tip where it contains the male inflorescence called tasel. Maize leaves have the typical grass leaf structure consisting of the main parts: the blade and the sheath surround the stalk. The leaves actematic between opposite sides of the stalk. Maize unlike other cereals has separate male and female flowers on the same plant. This implies that maize is monoecious plant. The male inflorescence, the tessel bearing staminate spikelet is produced at the apex of the main stem. Each female flower contains a single ovary terminates by a long style, the pollen-receptive organ commonly known as the silk. The silks are covered with fine hairs which trap wind-blown pollen. Each silk is a potential seed or kernel and must be pollinated for the seed to develop.

2.2. Growth requirements

Maize thrives under a wide range of environmental conditions, but for optimum production they require a lot of sunshine an warmth, average temperature below 14°C. Generally, large quantities of water well distributed are needed for high maize yields, maize uses water relatively efficiently. Most soils in Nigeria are acidic and are often deficient in essential plant nutrients (Ohiri and Ano, 1989). Maize requires a well aerated, deep, loamy, soils high inorganic matter, nitrogen, phosphorus and potassium. It can be grown successfully in soils whose pH ranges from 5.5 to 8.0. Since most soils in Nigeria is acidic, it can be improved using Biochar. Biochar not only increases soil pH but also affect the availability of other minerals and nutrients found in the soil (Mclean, 1982).

The optimum time of planting for early maize depends on the various ecological zones of Nigeria and was stated thus: forest zone (mid-March to first week of April), Sudan savanna (first to second week of June), Guinea Savanna (last week of May to first week of June), Southern Guinea Savanna (first last week of May to first week of May) Eze (2011). With the advent of global warming and climate change especially in Africa these schedules may not hold. Efforts should be geared up towards determination of time of planting for maize because of its sensitivity to this particular factor.

Complete fertilizer is applied at planting or immediately after germination. Compound fertilizer of nitrogen phosphorus and potassium, N.P.K (15: 15:15) at about 300-500 kg/hectare can be used. At about the time of tasselling, nitrogen fertilizer in the form of sulphate of ammonia $(NH_4)_2$ or Nitro Chalk $Ca(NO_3)_2$ can be placed about 5-10 cm away from the plant in ring or continuous rows. Where organic manure is available, it can be spread and work in before planting.

2.3. Maize production constraints

Maize suffers from parasitic and non-parasitic disease. The parasitic diseases are caused by fungi, bacteria, viruses, nematodes and parasitic seed plant. Maize is susceptible to a number of pests which attack it either in the field or storage. Major field pests of maize includes: stem borers, Army norms (Field Army Worms, FAW), earthworms, while storage pests includes; termite, rodents and weevils.

Weed is also a problem that affects the growth of maize. The limited use of organic and inorganic fertilizer and the declining soil fertility are problems to maize production in Sub-Saharan Africa. Periodic drought caused by irregular rainfall distribution reduces maize yield by an average of 15% each year, (Adams and Pearson, 1967).

Stalk rots are very destructive and in most cases the rots are caused by a combination of fungi and bacteria that attack plants approaching maturity. Its development is favored by dry weather early in the growing season.

Insects and birds damages to ears and seeds increase the incidence and seventy percent of rotting, all things being equal. Logged plants with the ears touching the ground expose the ears and kernels to more damage by insects and birds and consequently by rotting organisms.

Storage rots may develop on maize in cribs, bins or silos. If the moisture of the kernel (seeds) is above 12 or 15% and the surrounding air is warm enough for growth of fungi. It reduces market grade and feeding value.

2.4. Economic importance of maize

Worldwide, corn provides over 20 million tons of protein annually, used in production of alcohol (gin, whiskey, beer, industrial alcohol etc.) tooth paste, baby powder, cakes of soap, aspirin and other medicinal tablet may contain maize products, all kinds of paper must have starch treatment to give them a smooth surface and corn starch may be used for such treatment. Others are corn oil, corn flakes, corn meal flour, gums used in adhesives and sizing agents, syrup, dextrose, gluten man-power etc. Corn husks are used as filling materials whereas the corn cobs are used as fuel for making charcoal and for preparation of industrial solvents. Maize is used as feed grain and silage or green fodder for livestock. Maize consumption in Western Nigeria varies from 2.6 kg-2.8 kg per person per week (Owolabi *et al.*, 2003).

It is also used to make chicha, a fermented beverage of South America. It is also important in pediatrics health care delivery. The predominance of glycosides of linnobic acid and high glycosides content in the unsaturated acid, render it valuable in reducing serum cholesterol level in blood. The level which is associated with arterioderasis and coronary heart diseases.

Maize feedstock used for ethanol production, somewhat exceeded direct use for livestock fee. A fraction of the maize feedstock dry matter used for ethanol production is usefully recovered as Dried Distillers Grains with Solubles (DDGS), this DDGS are fed to livestock and poultry. Because starch utilization in fermentation for ethanol production leaves other grain constituents more concentrated in the residue, the feed value per kg of DDGS, with regard to ruminant-metabolizable energy and protein, exceeds that of the grain (Hoffman and Baker, 2011).

Maize is a major source of both grain feed and fodder for livestock. It is fed to the livestock in various ways. When it is used as a grain crop, the dried kernels are used as fed. Starch from maize can also be made intro plastics, fabrics, adhesives and many other chemical products. The corn sleep liquor, a plentiful watery by product of maize wet milling process, is widely used in the biochemical industry and research as a culture medium to grow many kind of microorganisms. Maize is also increasingly used as a feedstock for the production of ethanol fuel (biofuel), (Torres *et al.*, 2016)

2.5. Biochar: An overview

Biochar is a charcoal used as a soil ameliorant. It is recognized as a multifunctional material that can be used in carbon sequestration, metal immobilization and soil fertility increase. Biochar is a stable solid, rich in carbon and can endure in the soil for thousands of years. Like most charcoal, biochar is made from renewable materials via pyrolysis. The alkaline nature of biochar along with the presence of a variety of other elements depending on its origin makes it a good material as a soil conditioner.

Like most charcoal, biochar is made from biomass via pyrolysis. Biochar is under investigation as an approach to carbon sequestration. Biochar thus has the potential to help mitigate climate change via carbon sequestration. Independently, biochar can increase soil fertility of acidic soils (low pH soils), increase agricultural productivity, and provide protection against some foliar and soil-borne diseases. History The word "biochar" is a combination of "bio-" as in "biomass" and "char" as in "charcoal". It has been used in scientific literature of the 20[th]and 21st century. Pre-Columbian Amazonians are believed to have used biochar to enhance soil productivity. They seem to have produced it by smoldering agricultural waste (i.e., covering burning biomass with soil) in pits or trenches. European settlers called it terrapreta de Indio. Following observations and experiments, a research team working in French Guiana hypothesized that the Amazonian earthworm Pontoscolexcorethrurus was the main agent of fine powdering and incorporation of charcoal debris to the mineral soil.

2.6. Biochar production

Biochar is a high- carbon, fine-grained residue that today is produced through modern pyrolysis processes; it is the direct thermal decomposition of biomass in the absence of oxygen (Preventing combustion), which produces a mixture of solids (the biochar proper),liquid (bio- oil), and gas (synthesis gas) products. The specific yield from the pyrolysis is dependent on process condition, such as temperature, and can be optimized to produce either energy or biochar.

Temperatures of 400– 500 °C (673–773 K) produce more char, while temperatures above 700 °C (973 K) favor the yield of liquid and gasfuel components. Pyrolysis occurs more quickly at the higher temperatures, typically requiring seconds instead of hours. Typical yields are 60% bio-oil, 20% biochar, and 20% synthesis gas.

By comparison, slow pyrolysis can produce substantially more char (~50%); it is this which contributes to the observed soil fertility of terra preta. Once initialized, both processes produce unit energy. For typical inputs, the energy required to run a "fast" pyrolyzer is approximately 15% of the energy that it outputs. Modern pyrolysis plants can use the syntheticgas created by the pyrolysis process and output 3–9 times the amount of energy required to run.

The Amazonian pit/ trench method, harvests neither bio- oil nor synthetic gas, and releases a large amount of CO_2, black carbon, and other greenhouse gases (GHG)s (and potentially, toxins) into the air. Commercial-scale systems process agricultural waste, paper byproducts, and even municipal waste and typically eliminate these side effects by capturing and using the liquid and gas products. The production of biochar as an output is not a priority in most cases.

2.7. Centralized, decentralized, and mobile systems

In a centralized system, all biomass in a region is brought to a central plant for processing. Alternatively, each farmer or group of farmers can operate a lower-tech kiln. Finally, a truck equipped with a pyrolyzer can move from place to place to pyrolyze biomass. Vehicle power comes from the syntheticgas stream, while the biochar remains on the farm. The biofuel is sent to a refinery or storage site. Factors that influence the choice of system type include the cost of transportation of the liquid and solid byproducts, the amount of material to be processed, and the ability to feed directly into the power grid. For crops that are not exclusively for biochar production, the residue-to- product ratio (RPR) and the collection factor (CF) the percent of the residue not used for other things, measure the approximate amount of feed stock that can be obtained for pyrolysis after harvesting the primary product. For instance, Brazil harvests approximately 460 million tons (MT) of sugarcane annually, FAOSTAT (2006) with an RPR of 0.30, and a CF of 0.70 for the sugarcane tops, which normally are burned in the field. This translates into approximately 100 MT of residue annually, which could be pyrolyzed to create energy and soil additives. Adding in the biogases (sugarcane waste) (RPR=0.29 CF=1.0), which is otherwise burned (inefficiently) in boilers, raises the total to 230 MT of pyrolysis feedstock. Some plant residue, however, must remain on the soil to avoid increased costs and emissions from nitrogen fertilizers. Pyrolysis technologies for processing loose and leafy biomass produce both biochar and synthetic gas. Thermo catalytic depolymerization Alternatively, "thermo catalytic depoymerization which utilizes microwaves has recently been used to efficiently convert organic matter to biochar on an industrial scale, producing~50% char. Uses Carbon sink See also: Climate engineering The burning and natural decomposition of biomass and in particular agricultural waste adds large amounts of CO_2 to the atmosphere.

2.8. Uses of biochar

Biochar is a stable way of storing carbon in the ground for centuries, potentially reducing or stalling the growth in atmospheric greenhouse gas levels; at the same time its presence in the earth can improve water quality, increase soil fertility, raise agricultural productivity, and reduce pressure on old-growth forests. Biochar can sequester carbon in the soil for hundreds to thousands of years, like coal. Such a carbon-negative technology would lead to a net withdrawal of CO_2 from the atmosphere, while producing consumable energy.

This technique is advocated by prominent scientists such as James Hansen, head of the NASA Goddard Institute for Space Studies, and James Lovelock, creator of the Gaia hypothesis, for mitigation of global warming by greenhouse gas remediation. Researchers have estimated that sustainable use of biocharring could reduce the global net emissions of carbon dioxide (CO_2), methane, and nitrous oxide by up to 1.8 Pg CO_2-C equivalent (CO_2-Ce) per year (12% of current anthropogenic CO_2- Ce emissions; 1Pg=1 Gt), and total net emissions over the course of the next century by 130 Pg CO_2-Ce, without endangering food security, habitat, or soil conservation. Soil amendment biochar is recognized as offering a number of benefits for soil health. Many benefits area related to the extremely porous nature of biochar. This structure is found to be very effective at retaining both water and water-soluble nutrients. Soil biologist Elaine Ingham indicates the extreme suitability of biochar as a habitat for many beneficial soil micro organisms. She points out that when pre-charged with these beneficial organisms biochar becomes an extremely effective soil amendment promoting good soil, and in turn plant, health.

Biochar has also been shown to reduce leaching of E- coli through sandy soils depending on application rate, feedstock, pyrolysis temperature, soil moisture content, soil texture, and surface properties of the bacteria. For plants that require high potash and elevated pH, biochar can be used as a soil amendment to improve yield. Biochar can improve water quality, reduce soil emissions of green house gases, reduce nutrient leaching, reduce soil acidity, and reduce irrigation and fertilizer requirements. Biochar was also found under certain circumstances to induce plant systemic responses to foliar fungal diseases and to improve plant responses to diseases caused by soil borne pathogens. The various impacts of biochar can be dependent on the properties of the biochar, as well as the amount applied, and there is still a lack of knowledge about the important mechanisms and properties. Biochar impact may depend on regional conditions including soil type, soil condition (depleted or healthy), temperature, and humidity. Modest additions of biochar to soil reduce nitrous oxide N_2O emissions by up to 80%and eliminate methane emissions, which are both more potent greenhouse gases than CO_2. Studies have reported positive effects from biochar on crop production in degraded and nutrient–poor soils. The application of compost and biochar under FP7 project FERTIPLUS has had positive effects in soil humidity, and crop productivity and quality in different co untris.

Biochar can be designed with specific qualities to target distinct properties of soils. In a Columbian savanna soil, biochar reduced leaching of critical nutrients, created a higher crop uptake of nutrients, and provide greater soil availability of nutrients. At 10% levels biochar reduced contaminant levels in plants by up to 80%, while reducing total chlordane and DDX content in the plants by 68 and 79%, respectively. On the other hand, because of its high adsorption capacity, biochar may reduce the efficacy of soil applied pesticides that are needed for weed and pest control. High- surface-area biochars may be particularly problematic in this regard; more research into the long-term effects of biochar addition to soil is needed. Slash-and- char Switching from slash-and-burn to slash-and-char farming techniques in Brazil can decrease both deforestation of the Amazon basin and carbon dioxide emission, as well as increase crop yields. Slash-and-burn leaves only 3% of the carbon from the organic material in the soil. Slash-and-char can keep up to 50% of the carbon in a highly stable form. Returning the biochar into the soil rather than removing it all for energy production reduces the need for nitrogen fertilizers, thereby reducing cost and emissions from fertilizer production and transport. Additionally, by improving the soil's ability to be tilled, fertility, and productivity, biochar– enhanced soils can indefinitely sustain agricultural production, whereas non-enriched soils quickly become depleted of nutrients, forcing farmers to abandon the fields, producing a continuous slash and burn cycle and the continued loss of tropical rainforest. Using pyrolysis to produce bio-energy also has the added benefit of not requiring infrastructure changes the way processing biomass for cellulosic ethanol does.

Additionally, the biochar produced can be applied by the currently used machinery for tilling the soil or equipment used to apply fertilizer. Water retention Biochar is hygroscopic. Thus it is a desirable soil material in many locations due to its ability to attract and retain water. This is possible because of its porous structure and high surface area. As a result, nutrients, phosphorus, and agrochemicals are retained for the plants benefit. Plants are therefore healthier, and less fertilizer leaches into surface or groundwater.

2.9. Bio-oil and synthetic gas

Biofuel Mobile pyrolysis units can be used to lower the costs of transportation of the biomass if the biochar is returned to the soil and the syntheticgas stream is used to power the process. Bio-oil contains organic acids that are corrosive to steel containers, has a high water vapor content that is detrimental to ignition, and, unless carefully cleaned, contains some biochar particles which can block injectors. Currently, it is less suitable for use as a kind of biodiesel than other sources. If biochar is used for the production of energy rather than as a soil amendment, it can be directly substituted for any application that uses coal. Pyrolysis also may be the most cost-effective way of electricity generation from biomaterial.

2.10. Direct and indirect benefits of biochar

The pyrolysis of forest- or agriculture-derived biomass residue generates a biofuel without competition with crop production. Biochar is a pyrolysis byproduct that may be ploughed into soils in crop fields to enhance their fertility and stability, and for medium- to long-term carbon sequestration in these soils. It has meant a remarkable improvement in tropical soils showing positive effects in increasing soil fertility and in improving disease resistance in West European Soils. Biochar enhances the natural process: the biosphere captures CO_2, especially through plant production, but only a small portion is stably sequestered for a relatively long time (soil, wood,etc.). Biomass production to obtain biofuels and biochar for carbon sequestration in the soil is a carbon- negative process, i.e. more CO2is removed from the atmosphere than released, thus enabling long-term sequestration. Research Intensive research into manifold aspects involving the pyrolysis/biochar platform is underway around the world. From 2005 to 2012, there were 1,038 articles that included the word "biochar" or "bio-char" in the topic that had been indexed in the ISI Web of Science. Further research is in progress by such diverse institutions around the world as Cornell University, the University of Edinburgh, which has a dedicated research unit., the Agricultural Research Organization (ARO) of Israel, Volcanic Center, where a network of researchers involved in biochar research (IBRN, Israel Biochar Researchers Network) was established as early as 2009, and the University of Delaware. Long-term effect of biochar on soil C sequestration of recent carbon inputs has been examined using soil from arable fields in Belgium with charcoal-enriched black spots dating >150 years ago from historical charcoal production mound kilns.Topsoils from these 'black spots' had a higher organic C concentration [3.6 ± 0.9% organic carbon (OC)] than adjacent soils outside these black spots (2.1 ± 0.2% OC). The soils had been cropped with maize for at least 12 years which provided a continuous input of C with a C isotope signature (δ13C) −13.1, distinct from the δ13C of soil organic carbon (−27.4 ‰) and charcoal (−25.7 ‰) collected in the surrounding area. The isotope signatures in the soil revealed that maize- derived C concentration was significantly higher in charcoal-amended samples('black spots') than in adjacent unamended ones (0.44% vs. 0.31%; P =0.02). Top soils were subsequently collected as a gradient across two 'black spots' along with corresponding adjacent soils outside these black spots and soil respiration, and physical soil fractionation was conducted. Total soil respiration (130 days) was unaffected by charcoal, but the maize-derived C respiration per unit maize-derived OC in soil significantly decreased about half (P < 0.02) with increasing charcoal- derived C in soil.

Maize-derived C was proportionally present more in protected soil aggregates in the presence of charcoal. The lower specific mineralization and increased C sequestration of recent C with charcoal are attributed to a combination of physical protection, C saturation of microbial communities and, potentially, slightly higher annual primary production. Overall, this study provides evidence of the capacity of biochar to enhance C sequestration in soils through reduced C turn over on the long term. (Hernandez-Soriano *et al.*, 2015). Biochar sequesters carbon (C) in soils because of its prolonged residence time, ranging from several years to millennia. In addition, biochar can promote indirect C- sequestration by increasing crop yield while, potentially, reducing C-mineralization. Laboratory studies have evidenced effects of biochar on C-mineralization using 13C isotopes ignatures. (Kerre *et al.*, 2016) Fluorescence analysis of the dissolved organic matter from soil amended with biochar revealed that biochar application increased a humic- like fluorescent component, likely associated with biochar-carbon in solution. The combined spectroscopy-microscopy approach revealed the accumulation of aromatic-carbon indiscrete spots in the solid-phase of micro aggregates and its co-localization with clay minerals for soil amended with raw residue or biochar. The co- localization of aromatic- C polysaccharides C was consistently reduced upon biochar application. These finding suggested that reduced C metabolism is an important mechanism for C stabilization in biochar-amended soils (Hernandez-Soriano *et al.*, 2016) Students at Stevens Institute of Technology in New Jersey are developing super capacitors that use electrodes made of biochar. A process developed by University of Florida researchers that remove phosphate from water, also yields methane gas usable as fuel and phosphate-laden carbon suitable for enriching soil.

2.11. Possible commercial use

If biomass is pyrolyzed to biochar and put back into the soil, rather than being completely burned, this may reduce carbon emissions. Potentially, the bio energy industry might even be made to sequester net carbon. (Lehmann, 2007) Pyrolysis might be cost-effective for a combination of sequestration and energy production when the cost of CO_2 ton reaches \$37. Current biochar projects make no significant impact on the overall global carbon budget, although expansion of this technique has been advocated as a geo engineering approach. In May2009, the Biochar Fund, a small "social profit organization", received a grant from the Congo Basin Forest Fund for a project in Central Africa to simultaneously slow down deforestation, increase the food security of rural communities, provide renewable energy and sequester carbon. Though some farmers did report better maize crops, the project ended early without significant results and with promises to the farmers not kept. Application rates of 2.5–20 tons per hectare (1.0–8.1 tons/acre) appear to be required to produce significant improvements in plant yields. Biochar costs in developed countries vary from $300–7000/ton, generally too high for the farmer/horticulturalist and prohibitive for low- input field crops. In developing countries, constraints on agricultural biochar relate more to biomass availability and production time. An alternative is to use small amounts of biochar in lower cost biochar-fertilizer complexes. Various companies in North America, Australia, and England sell biochar or biochar production units. In Sweden the 'Stockholm Solution' is an urban tree planting system that uses 30%biochar to support healthy growth of the urban forest.

The Qatar Aspire Park now uses biochar to help trees cope with the intense heat of their summers. At the 2009 International Biochar Conference, a mobile pyrolysis unit with a specified intake of 1,000 pounds (450 kg) was introduced for agricultural applications. The unit had a length of 12 feet and height of 7 feet (3.6 m by 2.1m). A production unit in Dunlap, Tennessee by Mantria Corporation opened in August 2009 after testing and an initial run, was later shut down as part of a Ponzi scheme investigation.

2.12. The effect of biochar on the physical and chemical properties of the soil

The physical and chemical properties of biochar are keys to understand performance and mechanisms of biochar in the improvement of soil fertility. A possible main mechanisms for yield improvement may be the increase of soil water holding capacity after biochar treatment, Jeffrey *et al.* (2011). Biochar has high total porosity and it could both retain water in small pores and thus increase water holding capacity and assist water to infiltrate from the ground surface to the topsoil through the larger pores after heavy rainfall, Asai (2009).

Peake *et al.* (2014) indicated that biochar application could increase available water capacity by over 22%. Neliseen *et al.* (2015) demonstrated that biochar application could increase available water capacity from 0.12 -0.13 m3m-3. Moreover, the formation and stability of soil aggregates could increase the crop production and the prevention of soil degradation, Amazketa (1999). The capacity of soil aggregation increased ranging from 8-36% after biochar treatment from rice husk, Lusg *et al.* (2014). They also reported that the application of rice mill waste biochar could increase soil pore structure parameters by 20% and shear strength, as well as decrease soil swelling by 11.1%. in addition, biochar could ameliorate compaction by over 10%, decrease bulk density from 1.47-1.44 mg m-3 and increase porosity from 0.43-0.44m3 m-3 Neliseen *et al.* (2015). Overall, the improved physical properties of the soil such as; bulk density, water holding capacity and aggregate stability may increase the retention of both water and nutrient which benefits the soil fertility directly. The application of biochar could increase soil pH.

Weng *et al.* (2014) reported that rice mill waaste biochar increased the total tea garden soil (acidic soil) pH from 3.33-3.36. The Agricultural soil pH increased by almost IPH unit for biochar treatment which is produced from mixed hardwood (Querus spp. And Carya spp). Laird *et al.* (2010). The increase of soil pH could change the form of nutrient and facilitate some elements adsorption of the root. Cation Exchange Capacity (CEC) is the indirect measures of the capacity of soils to retain water and nutrients, indicated that the biochar treatments significantly increased CEC by 4-30% and relative to the controls. Similarly, CEC of the highly leathered soils was increased from 7.41 to 10.8 Cmol kg^1 after biochar treatment, which is produced from *Leucaemaleucocephala,* J in and Wang (2013).

In addition, biochar treatment can increase base saturation percentage from 6.4-26% and saturated hydraulic conductivity from 16.7-32.1 Cmh^{-1} and increase total C from 2.27-2.78, total N from 0.24-0.25, available P from 15.7-15.8 mgkg^{-1} Jones *et al.* (2012).

Overall, the improvement of soil properties and directly or indirectly increase nutrient content and availability and decrease nutrient leaching which are known as mechanism for the increase of soil fertility.

3. MATERIALS AND METHODS

3.1. Description of the experimental site

The experiment was carried out in the 2018 planting season at the Faculty of Agricultural and Natural Resources, Teaching and Research Farm Enugu State University of Science and Technology (ESUT), Enugu, Nigeria, (latitude $06^{\circ}52^{1}N$ and longitude $07^{\circ}15^{1}E$). The area has an annual rainfall of 1700-2010mm. the rainfall pattern is bimodal between April and October and dry between November and March. Anikwe *et al.* (2005)

3.2. Source of materials

The materials used include:

i)   Biochar from rice mill waste; this was gotten from Ugbawka rice mill site, Nkanu East Local Government Area, Enugu State.

ii)   Maize (Oba supper II) was sourced from Eke Agbani Market in Nkanu West Local Government Area, Enugu State.

3.3. Field experimental layout

The experimental layout was Randomized Complete Block Design with four treatments and five replications.

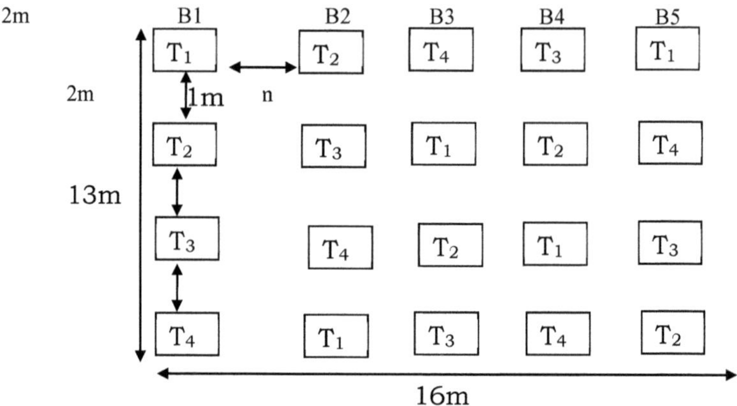

Figure1: Field layout

B: Block

Treatment rates:

$T_1 = 0$ t/ha

$T_2 = 10$ t/ha

$T_3 = 20$ t/ha

$T = 30$ t/ha

The experimental site was mapped out and was cleared of existing grasses with the aid of a cutlass. A total land area of 13m x 16m (208 m²) was mapped out with a measuring tape and pegs. The field was divided into five blocks and four experimental units making a total of 20 plots. The spacing between the blocks was one meter while the distance between the replicates was one meter. Soil samples from (0-35) cm was collected and taken for laboratory analysis before planting. The treatment was applied at the rates of $T_1 = 0$ t/ha, $T_2 = 10$ t/ha, $T_3 = 20$ t/ha, $T_4 = 30$ t/ha. Rice mill waste (husk) was used to produce biochar. The waste was collected from Ugbawka Rice Milling Site. The waste was sun dried then underwent pyrolysis under limited oxygen. Carbonization was performed at 7500°C and was held for two hours before cooling at room temperature. The biochar was then stored in air tight containers.

The treatment was replicated five times in a Randomized Completed Block Design (RCBD) each of the plot size was 2m x 2m, the treatment was incorporated into the soil during tillage.

Maize grains was planted two seeds per hole at a spacing of 25 x 75cm and thinning was done to one stand seven days after germination. Manual weeding was carried out 4 and 8 weeks after planting.

3.4. Data collection

The data collected were: plant height (this was done by measuring the length of four plants from the center leaving the first row which served as my sampling unit), the leave area index was measured from the leaves of the same sampling unit by the:

$$\frac{\text{No. of leaves} + (\text{leave width} \times \text{leave length})}{\text{Spacing distant}} \quad \ldots\ldots\ldots \text{ equation 1}$$

90 days after planting, percentage emergence one week after planting was calculated by counting the number of germinated seeds divided by the number of plant x 100, and the number of cobs.

A composite soil-sample was collected before planting with the aid of an auger within the depth of (0-35cm) from each plot and another sample was collected after 90 days. The samples was dried and bulked for analysis. The samples were analyzed for: total N (Nitrogen) using Micro Kjeldal Method (Bremmer, 1982), available Phosphorous (P), Potassium (K), Carbon (C), Calcium (Ca), Magnesium (Mg), sodium (Na), Cation Exchange Capacity (CEC) and soil pH. Available P was determined using Bray 11 method (Bray and Kurtz, 1945) soil pH in H_2O was measured by electrometric method (Mclean, 1982). The Cation Exchange Capacity (CEC) was determined by the method described by Thomas (1982).

3.4. Data analysis

The data collected from the experiment were analyzed using analysis of variance test for randomized complete block design according to procedures outlined by Steel and Torrie (1980). Significant means were separated using Fishers least significant different at 5% probability level as described by Obi (2002).

Table outline of the Analysis of Variance (ANOVA)

Source of variation	Degree of freedom (general)	Degree of freedom (specific)
Block	r-1	4
Treatment	t-1	3
Error	(r-1)(t-1)	12
Total	rt-1	19

4. RESULTS AND DISCUSSION

The result of the soil analysis shown in table one indicated that the textural class of the soil was loamy sand USDA (1993). It contained clay particle size of 6%, silt 7% and sand 45% with pH 5.0 and 4.6 in H_2O and KCl as extractants respectively. Soil with this pH range, generally have a low availability of calcium, magnesium and phosphorous. Anikwe *et al.* (2005).pH determination in the soil water suspension normally provides adequate information and has the merit of simplicity (Ladon, 1991).

Available phosphorous was at the range of 8.52 Mg kg^{-1} but increase to 10.25 Mgkg^{-1} after the treatment. Organic matter content was 0.64 % before the biochar was added but increased to 1.10 %. There was significant effect in the amount of total Nitrogen from 0.007 % to 0.063 %. The CEC value increased after the treatment and the value of calcium (Ca) and Magnesium (Mg) increased from 0.70 C mol kg-^{1} to 1.62Cmol kg-^{1} and 1.50Cmol kg-^{1} to 1.70 C mol kg-^{1} respectively while Sodium (Na) and Potassium (K) increased also from 0.02 C mol kg-^{1} to 0.03Cmol kg-^{1} and 0.03 C mol kg-^{1} to 0.07 C mol kg^1 respectively. The base saturation, hydrogen and carbon increased from 35.71 % to 38.64 %, 1.00 C mol kg-^{1} to 1.30 C mol kg-^{1} and 4.6 % to 5.7 % respectively.

The statistical analysis revealed a significant difference in pH among the treatments due to the addition of biochar. The highest value was observed in the soil treated with 30 t/ha while the lowest value was observed at the control 0 t/ha. There was an increase in soil pH value from 5.3 to 5.3 and 5.7 at 10, 20 and 30 t/ha respectively. The increase in soil pH was generally attributed to ash accretion as ash residues are generally dominated by carbonate of alkali and alkaline earth metal and a considerable amount of silica. Lehmann *et al.* (2009). The high surface area and porous nature of biochar may be another reason for the increase in soil pH after application, by increasing the Cation Exchange Capacity, Nigussie *et al.* (2012).

The results in table two also indicated that rice waste biochar can be used as a substitution for liming to increase the pH of acidic soil, Masulili *et al.* (2010). The application of rice waste biochar also increased elemental plant nutrients like Nitrogen and

Phosphorus for all treatment, 10, 20 and 30 t/ha compared to the control 0 t/ha. Similar results were reported by Glaser (1999). It also increased the Cation Exchange Capacity 6.20 C mol kg-^{1} at 10t/ha, 7.4 C mol kg-^{1} at 20 t/ha and 8.72 C mol kg-^{1} at 30 t/ha.

Table1: The soil physiochemical properties before planting (Pre-Planting)

Parameter	Value
Clay (%)	6
Silt (%)	7
Fine sand (%)	45
Coarse sand (%)	42
Textural Class	Loamy sand
pH (H$_2$O)	5.0
(KCl)	4.6
Carbon (%)	0.403
Organic matter (%)	0.642
Nitrogen (%)	0.007
Sodium (C mol kg-^{1})	0.02
Potasium (C mol kg-^{1})	0.03
Calcium (C mol kg-^{1})	0.70
Magnesium (C mol kg-^{1})	1.50
Cation Exchange Capacity (C mol kg-^{1})	6.20
Base Saturation. (%)	35.71
Hydrogen (C mol kg-^{1})	1.00
Available Phosphorus (Mgkg^{-1})	8.52

Table 2: The effect of biochar on Cation Exchange Capacity (CEC), pH, Total Nitrogen and Available Phosphorus

Treatment rate (t/ha)	CEC	pH	Nitrogen	Available phosphorus
0	8.71	4.9	0.006	8.50
10	6.20	5.2	0.008	8.55
20	7.41	5.3	0.063	9.21
30	8.72	5.7	0.068	10.26
FLSD $_{(0.05)}$	**0.02**	**0.01**	**0.02**	**0.4**

FLSD$_{(0.05)}$ – Fishers Least Significant Difference at 5% level probability

Effect of different rate of rice mill waste biochar on percentage emergence one week after planting

The result in table three showed that the treatment affected the plant emergence but not according to the exact rate of application. The highest percentage of emergence was found in T_3 (97.3%) which has the rate of 20 t/ha differing from the T_4 with a rate of 30 t/ha. this might be due to some climatic factors.

Tracy *et al.* (2018) reported that increase in the productivity of crop planted on soil treated solely with biochar should not be expected, as that biochar is not a fertilizer but a soil ameliorant.

Table 3: Effect of different rate of rice mill waste biochar on percentage Emergence one WAP

Treatment rate (t/ha)	Percentage emergence (%)
0	95.00
10	96.20
20	97.30
30	95.00
FLSD$_{(0.05)}$ for any 2 rates	**0.09**

FLSD$_{(0.05)}$ Fishers Least Significance Difference at 5% level probability

WAP- Week After Planting

The effect of different rate of rice mill waste biochar on plant height.

The result in table four showed that the treatment also did not affect the plant height according to the rate of application. This maybe also due to some climatic or edaphic factors. The highest plant was found in T_3; 5.0 with height rate of 156.2 cm, followed by the 30 t/ha rate with a height of 153.3 cm T_4, the T_1 (0 t/ha) and the T_2; 10 t/ha

An increase in the rate of application of biochar, will also increase the response of maize to the uptake of nutrients available if the above factors are favorable.

Table 4: The effect of different rate of rice mill waste biochar on plant height 90 days after planting.

Treatment rate (t/ha)	Plant height (cm)
0	152.20
10	141.20
20	156.20
30	153.30
FLSD $_{(0.05)}$ for any 2 rates	**1.10**

F-LSD$_{(0.05)}$ – Fishers Least Significant Difference at 5% level probability

DAP – Days After Planting

The effect of different rates of rice mill waste biochar on plant number of cobs per plant.

Table five result showed that the treatment affected the plant number of cobs. The highest mean number of cobs was found in plots treated with T4 (30 t/ha), compared to the other treatment rates which had a value of 1.35. The treatment did not affect the number of cobs in the other rates as applied; this may be due to climatic or edaphic factors.

The number of cobs is an important yield contributing parameter. It contributes to the yield of maize by influencing both numbers of grains per cob and grain size. From the result, an increased application is suitable for the rice waste biochar to reduce the soil acidity, thereby keeping it in an improved condition.

Table 5: The effect of different rates of rice mill waste biochar on plant number of cobs per plant 90 DAP.

Treatment rate (t/ha)	Number of cobs per plant
0	1.25
10	1.25
20	1.15
30	1.35
FLSD $_{(0.05)}$ for any 2 rates	**0.02**

FLSD$_{(0.05)}$ – Fishers Least Significant Difference at 5% level probability

DAP – Days After Planting

Effect of different rate of rice mill waste biochar on leaf area index per plant 90 days after planting

The result in table six showed that the treatment did not have an effect on the leaf area index at 90 DAP. The highest mean of leaf area index was found in plots treated with T_1 (0 t/ha) with a value of 0.277, T_2 (10 t/ha) had 0.222, T_3 (20 t/ha) had 0.215 while the T_4 (30 t/ha) with the highest rate, had the lowest leaf area index value.

The result showed that the rice meal waste biochar is not profitable to the leaf area index of maize. The leaf area index is an important parameter in plant ecology, because it tells how much foliage there is, it is a measure of the photosynthetic active area, and at the same time of the area subjected to transpiration. It is also the area which becomes in contact to air pollutants. For the benefit of the maize growth, the photosynthetic reaction must be enhanced by the addition of organic or inorganic manure, since biochar is not a fertilizer but a soil amendment material.

Table 6: Effect of different rate of rice mill waste biochar on leaf area index 90 DAP

Treatment rate (t/ha)	Leaf Area Index per plant
0	0.28
10	0.27
20	0.22
30	0.21
$FLSD_{(0.05)}$ for any 2 rates	**0.02**

$FLSD_{(0.05)}$Fishers Least Significant Difference at 5% level probability

DAP – Days After Planting

Effect of different rate of rice mill waste biochar on weight of cobs per plant 90 DAP

The result in table seven showed that there was no effect of the treatment on the weight of cobs. The highest weight of cobs was found in plots treated with T_1: 0 t/ha followed by the T_4 30 t/ha and T_2 10 t/ha which are the same cob weight, while T_3 20 t/ha has the lowest weight.

This also showed that biochar cannot implement growth mechanism because it lacks elemental nutrient, but contains alkaline which serves as a soil conditioner.

Table 7: Effect of different rate of rice mill waste biochar on weight of cobs per plant 90 DAP

Treatment rate (t/ha)	Weight of cobs (kg)
0	0.32
10	0.32
20	0.29
30	0.32
$FLSD_{(0.05)}$ for any 2 rates	**0.01**

$FLSD_{(0.05)}$ – Fishers Least Significant Difference at 5% level probability

DAP – Days After Planting

Effect of different rates of rice mill waste biochar on the number of leaves per plant

The result in table eight showed that the treatment had an effect the number of leaves in the plots treated with T_4: 30 t/ha, while it varies between T_3: 20 t/ha, T_2: 10 t/ha and T_1: 0 t/ha.

To maintain a sustainable productivity, therefore, an addition of inorganic fertilizer may be necessary since biochar is not a fertilizer but an ameliorant, Tracy *et al.* (2018)

Table 8: Effect of different rates of rice mill waste on the number of leaves.

Treatment rate (t/ha)	Number of leaves per plant
0	11.00
10	10.45
20	10.85
30	11.35
$FLSD_{(0.05)}$ for any 2 rates	**0.02**

$FLSD_{(0.05)}$ – Fishers Least Significant Difference at 5% level probability

DAP – Days After Planting

5. CONCLUSION AND RECOMENDATION

The results of the study showed that the treatment (0 t/ha, 10 t/ha, 20 t/ha and 30 t/ha) of biochar had effect on the plant but not on all parameter and not according to the rate of application, this may be due to some edaphic or climatic factors. The study also showed that the treatments, 10 t/ha, 20 t/ha and 30 t/ha had some positive effect on the properties of the soil.

On the basis of this study the application of biochar at 30 t/ha was the most effective rate. Therefore, it is advisable that local farmers should make use of biochar derived from rice mill waste as it can be used as a substitute for lime material to increase the pH of acidic soils for the production of maize.

REFERENCES

Awad, Y.M., Blagodatskaya E., Y.S., Kuzyakov Y., (2012). Effects of polyacrylamide, biopolymer and biochar on decomposition of soil organic matter and plants residu es as determined by 14C and enzyme activities. Eur J Soil Biol 48:1–10.

Brady, N.C., and R.R., Weli (1999). The nature and properties of soils 12th edition. Prentice Hall.New Jersey pp.362

Ball, D.F., (1964) Loss- on-ignition as an estimate of organic matter and organic carbon in non- calcareous soil. J. Soil Sci. 15:84-92.

Blamey, F.P.C., Chapman, J., (1982) Soil amelioration effects on peanut growth, yield and quality. Plant Soil 65:319–334.

Bray, R.H., and Kurtz L.T., (1945). Determination of total organic and available forms of phosphate in soil. Soil Science 59:39-45.

Bremner, J.M., (1982). Total nitrogen in page et al (edg). Methods of soil analysis part 2 ASA No. 9. Madison Wisconsin. pp 595-624.

Chen, B., Chen Z., Lu S., (2011) A novel of magnetic biochar efficiently sorbs organic pollutants and phosphate. Bioresour Technology 102:716–723.

FAOSTAT. (2006) Production Quantity of Sugar Cane In Brazil In 2006".

Fitz Patrick, E.A., (1983) Soils: their formation, classification and distribution. Longman Science & Technical, London, p. 353.

Glaser, B., Lehmann, J. and Zech, W. (2002) Ameliorating physical and chemical properties of highly weathered soils in the Tropics with charcoal. A review.

Hernandez-Soriano, M.C Kerre, B. Goos, P. Hardy, B. Dufey, J. Smolders, E. (2015). Long-term effect of biochar on the stabilization of recent carbon:soils with historical inputs of charcoal" . GCB Bioenergy. 8 (2):371–381.

Hernandez- Soriano, M.C., Kerre, B., Kopittke, P., Horemans, B., Smolders, E. (2016). "Biochar effects on carbon composition and stability in soil: a combined microsco py study". Scientific Reports.6: 25127. Bibcode: 2016 NatSR...6251H doi:10.103 8/ srep25127 . PMC 4844975. PMID 27113269.

International Institute for Tropical Agriculture (1982) Maize production manual.Vol. 1. Manual series 8:1–2.

Jeffery S., Verheigen F.G.A., Van Der Valde, M., Bastos, A.C., (2011) A quantitative review of the effects of biochar application to soils on crop productivity using meta analysis. Agric Ecosystem Environ 144:175-187

Jones D.L., Rousk J., Edward-Jones G., Delues T.H., Murphy D.V., (2012). Biochar-mediated changes in soil quality and plant growth in a three year field trial. Soil Biochem 4.113-124

Kayode, G.O and A.A Agboola (1993). Investigation on the use of Macro and micro nutrients to improve maize yield in Southeastern Nigeria. Liming Research 4:211-121.

Kerre, B., Hernandez- Soriano, M.C., Smolders, E. (2016)."Partitioning of carbon sources among functional pools to investigate short-term liming effects of biochar in soil: a 13C study" . Science of the Total Environment. 547: 30–38. Bibcode: 2016ScTE n.547...30K .doi:10.1016/ j.scitotenv. 2015.12 .107 . PMID 26780129 . External links International Biochar Initiative. Biochar.org

Land D.A., Fleming P., Davis D.D., Horton R., Wang B., Karlen D.L., (2010) Impact of biochar amendments on the quality of a typical Midwestern Agricultural soil. Geoderma 158-443-44

Lehmann, J., Czimczik, C., Laird, D., and Sohi, S. (2009) Stability of biochar in the soil. In J. Lehmann and S Joseph, biochar for Environmental Management: Science and Technology. (pp: 183-205) London: Earthscan Publishing.

Lusg S., Zong Y.T. (2014) Effect of rice husk biochar and coal fly ash on some physical properties of expansive clayey soil (vertisol) catena 114:37-44

Masulili, A., Utomo, W. H. and Sychfari. M. (2010) Rice based biochar for rice based cropping system in acidic soil. Characteristics of rice waste biochar and its influence on the properties of acidic sulphate soil and rice growth, Kalimanta, Indonesia. Journal of Agricultural Science (pp: 39-47)

Moon, D.H., Chang Y.Y., Cheong K.H., Koutsospyros A., Park J.H., (2014a) Amelioration of acidic soil using various renewable waste resources. Environ SciPollut Res 21:774–780.

Nelissen V, Saha BK, Ruysschaert G, Boeckx p, (2014) Effect of different biochar and fertilizer types on N_2O and No emission. Soil Biol. Biochem 70:244-255

Nigussie, A., Kissi, E., Misganaw, M.andAmbaw, G. (2012) Effect of biochar Application on soil properties and Nutrient uptake in Lettuces. American Eurasian Agricultural and Environmental Sci. (pp: 369-376)

Obi I.U., 2002. Statistical methods of detecting difference between treatment means and research methodology issues in laboratory and field experiment.

Ohiri, A.C., and Ano A.O., (1989). Characterization and evaluation of some soils of rainforest zone of Nigeria.Proceeding 17th Annual conference of soil science society of Nigeria.Nsukka, `pp. 56-60.

Steel, G.D., and Torrie J.H., (1980).Procedures of statistics.A biometrical Approach 2nd edition.New York Mcgraw hHill box col pernyInc, 633p.

Tel, D.A., (1984) Soil and plant analysis. Department of land resources Science. University of Guelph, pp:277.

APPENDICES

APPENDIX I

Analysis of variance

Variate: PERCENTAGE EMERGENCE

Source of variation	d.f.	s.s.	m.s.	v.r.
REPLICATION	4	132.09	33.02	1.14
BIOCHARRATES	3	19.05	6.35	0.22
Residual	12	346.16	28.85	
Total	19	497.30		

Tables of means

Variate: PERCENTAGE EMERGENCE

Grand mean 95.9

BIOCHAR RATES	0.0	2.5	5.0	7.5
	95.0	96.2	97.3	95.0

Least significant differences of means (5% level)

Table	BIOCHAR RATES
rep.	5
d.f.	12
l.s.d.	7.40

APPENDIX II

Analysis of variance

Variate: LAI

Source of variation	d.f.	s.s.	m.s.	v.r.
REPLICATION	4	0.03739	0.00935	0.92
BIOCHAR RATES	3	0.01532	0.00511	0.50
Residual	12	0.12258	0.01021	
Total	19	0.17529		

Tables of means

Variate: LAI

Grand mean 0.230

BIOCHAR RATES	0.0	2.5	5.0	7.5
	0.277	0.222	0.215	0.206

Least significant differences of means (5% level)

Table	BIOCHAR_RATES
rep.	5
d.f.	12
l.s.d.	0.1393

APPENDIX III

Analysis of variance

Variate: NOC

Source of variation	d.f.	s.s.	m.s.	v.r.
REPLICATION	4	0.28125	0.07031	2.29
BIOCHAR RATES	3	0.10000	0.03333	1.08
Residual	12	0.36875	0.03073	
Total	19	0.75000		

Tables of means

Variate: NOC

Grand mean 1.250

BIOCHAR RATES	0.0	2.5	5.0	7.5
	1.250	1.250	1.150	1.350

Least significant differences of means (5% level)

Table	BIOCHAR_RATES
rep.	5
d.f.	12
l.s.d.	0.2416

APPENDIX IV
Analysis of variance

Variate: NOL

Source of variation	d.f.	s.s.	m.s.	v.r.
REPLICATION	4	2.331	0.583	0.50
BIOCHAR RATES	3	2.084	0.695	0.59
Residual	12	14.119	1.177	
Total	19	18.534		

Tables of means

Variate: NOL

Grand mean 10.91

BIOCHAR RATES   0.0   2.5   5.0   7.5

               11.00   10.45   10.85   11.35

Least significant differences of means (5% level)

Table	BIOCHAR_RATES
rep.	5
d.f.	12
l.s.d.	1.495

APPENDIX V

Analysis of variance

Variate: Plant HEIGHT

Source of variation	d.f.	s.s.	m.s.	v.r.
REPLICATION	4	3647.9	912.0	1.13
BIOCHAR RATES	3	643.5	214.5	0.27
Residual	12	9679.1	806.6	
Total	19	13970.5		

Tables of means

Variate: Plant HEIGHT

Grand mean 150.8

BIOCHAR RATES	0.0	2.5	5.0	7.5
	152.2	141.2	156.2	153.3

Least significant differences of means (5% level)

Table	BIOCHAR_RATES
rep.	5
d.f.	12
l.s.d.	39.14

APPENDIX VI

Analysis of variance

Variate: Weight of Cobs kg

Source of variation	d.f.	s.s.	m.s.	v.r.
REPLICATION	4	0.018312	0.004578	1.03
BIOCHAR RATES	3	0.003844	0.001281	0.29
Residual	12	0.053187	0.004432	
Total	19	0.075344		

Tables of means

Variate: Weight of Cobs _kg

Grand mean 0.309

BIOCHAR RATES	0.0	2.5	5.0	7.5
	0.320	0.315	0.285	0.315

Least significant differences of means (5% level)

Table	BIOCHAR RATES
rep.	5
d.f.	12
l.s.d.	0.0917

YOUR KNOWLEDGE HAS VALUE

- We will publish your bachelor's and
  master's thesis, essays and papers

- Your own eBook and book -
  sold worldwide in all relevant shops

- Earn money with each sale

Upload your text at www.GRIN.com
and publish for free